BASIC CHEMISTRY (1)

Preface

This book is genuinely written for grasping the fundamental concept of chemistry. It is aimed to the secondary level students. It can serve as a reference for a particular topic. It is also useful for various competitions.

Index

Introduction

Necessity is the mother of invention. Human curiosity and imagination had added a wing to our flight of invention. Science has no destination but journey. As a domain of science chemistry also deals with what? Why? How? From what the matter around us made up of? And what are the properties of these matters? These are the subjects of chemistry. It has a golden history from alchemists to the detection of higgs boson (commonly referred by journalists "the god particle"). It is assumed that alchemists used to convert other elements into gold. Now we are observing the same thing only the difference between two elements is the number of fundamental particles electron, proton, and neutrons. Now, we are aware of that the whole universe is made up of 118 elements (discovered till now). Our body, the daily items we use, medicine we take, and our surroundings all are the result of these elements only.

CHAPTER 0

SOME ELEMENTRY CONCEPTS

ELEMENTS: according to the Indian Vedas the whole universe is made up of five fundamental sources (building blocks). These are soil, water, fire, sky and air. As per our curiosity and experiment, we verified that these are not ultimate sources. Water is the combination of hydrogen and oxygen. Air is a mixture of many gases like nitrogen, hydrogen, chlorine etc. These basic building blocks are referred as elements. Till now we have 92 natural occurring elements and 26 synthesized in labs.

First 20 elements with their atomic no., atomic weight, and symbol.

atomic no.(z)	name of the elements	symbols	atomic mass(amu)
1	hydrogen	H	1
2	Helium	He	4
3	Lithium	Li	7
4	Beryllium	Be	8
5	boron	B	11
6	Carbon	C	12
7	Nitrogen	N	14
8	Oxygen	O	16
9	Fluorine	Ne	19
10	Neon	F	20
11	Sodium	Na	23
12	Magnesium	Mg	24
13	aluminum	Al	27
14	Silicon	Si	28
15	Phosphorus	P	31
16	Sulphur	S	32
17	Chlorine	Cl	35
18	Argon	Ar	40
19	Potassium	K	40
20	Calcium	Ca	40

Table 1

Some other elements and their symbols

Ag argentum (silver)

Au aurum (gold)

Hg hydrergyrum (mercury)

Pb plumbum (lead)

ATOM: the word "atom" is derived from the Latin word atomos which means uncut able. In India around 500 B.C maharishi Kanad gave the word parmanu which is now referred as atom. Each element has its unique atom from which it is made up of.

ELECTRONIC CONFIGURATION: Each atom has nucleus which consists of neutron and proton. Electrons revolve around the nucleus in specific elliptical orbits. Electronic configuration is the postal address of the electrons in an atom. In chemistry, electronic configuration is the basic tool through which we can find the desired property of an element. We can find the states of the element, valency of the atom. These are helpful in choosing a particular element for a particular purpose.

HOW TO FIND VALENCY FROM ELECTRONIC CONFIGURATION:

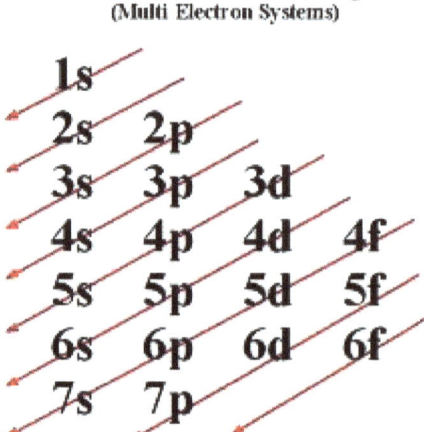

Fig. 1 filling of electrons in orbits

We can find the valency of a particular element in simple steps. Before using these steps we must know that the filling of electron is started from 1s then 2s and go on succession with 2p, 3s, 3p, 4s, 3d, 4p, 5s, 4d, 5p, 6s, 4f, 5d, 6p, 7s, 5f, 6d, 7p, 6f and so on as shown in fig. 1. Secondly we must also care that the max no. of electron in a particular shell like s, p, d, and f is limited i.e.

Shell	no. of max electron allowed
S	2
P	6
d	10
f	14

Working step:

Example 0.1: find the valency of sodium?

Solution:

Step: 1 find the electronic configuration

The atomic no. of sodium is 11. So, there are 11 electrons in its neutral atom.

Therefore filling of electrons will be as follows $1s^2$, $2s^2$, $2p^6$, $3s^1$.

Note: here we find that the max no. of electron can be 2 in 3s shells but already 10 electrons have been filled in 1s, 2s and 2p, so only 1 electron is filled in 3s.

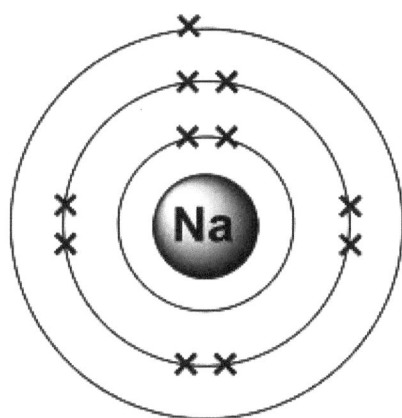

Fig.2 electronic configuration of Na

Step: 2 in the outer most shell there is only one electron. Hence its valency is one.

Note: valence electron refers to the no. of electrons residing in the outer most shell and only these electrons take part in a particular reaction. If there are more than 4 electron in the outer most orbit. For finding valency we must subtract the no of electrons from 8 (as it is necessary to complete octet). But this time valency is in negative due to receiving electron from other atom electron has negative charge.

We can see it in the next example:

Example 0.2: find the valency of oxygen?

Solution:

Step: 1 find the electronic configuration

The atomic no. of oxygen is 8. So, there are 8 electrons in its neutral atom.

Therefore filling of electrons will be as follows 1s^2, 2s^2, 2p^4.

Note: Here we find that the max no. of electron can be 6 in 2p shell but already 4 electrons have been filled in 1s and 2s so only 4 electrons is filled in 2p.

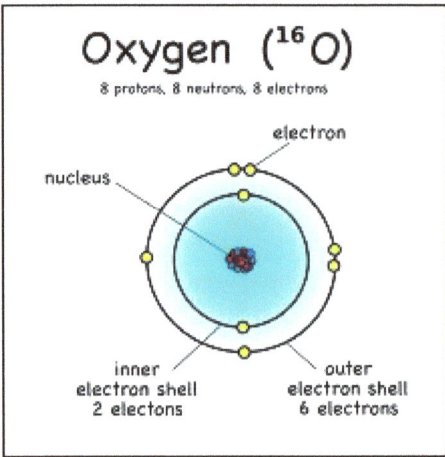

Fig.3 electronic configuration of Oxygen

Step: 2 in the outer most shell there are 6 electrons (2 in 2s and 4 in 2p as outer most orbit is 2) only , which is greater than 4. Hence its valency 6-8=-2.

CHEMICAL REACTIONS USING VALENCY:

Why does reaction take place?

It is general question, why does any reaction takes place? When H_2 and O_2 combine only H_2O is formed not other compound? Why does H_2 Combine with O_2 but not with He? Every atom wants to get a stable configuration. Any atom having full filled outer orbit is considered as a stable atom. An atom can have 8 electrons in its outer most orbits to get its stable condition referred as octet. In case of H and He, it is 2 and referred as duplet. When two or more atoms come in vicinity or contact they exchange their electron so as to get a stable configuration. The combining capacity of an atom is referred as valency.

For example: when Na and Cl react, Sodium has 1 electron in its outer orbit and chlorine has 7. If Na loses its 1 electron and chlorine gains this both will have a stable configuration. In this process Na changed to +ve ion and Chlorine to –ve. Now ionic bonding makes it possible to form NaCl.

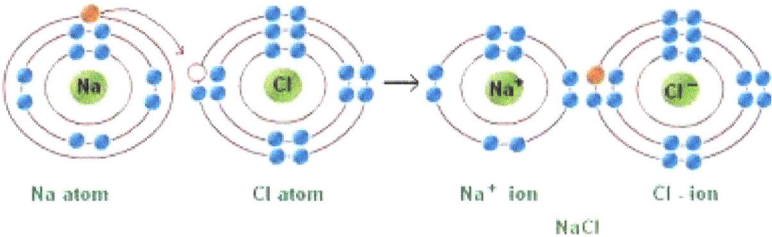

Fig 4. Formation of NaCl

In the above example we see that the actual process is the swapping or exchange of valencies. If we know the valences' we can easily write the chemical formula. Following diagrams will explain it clearly.

Example 0.3

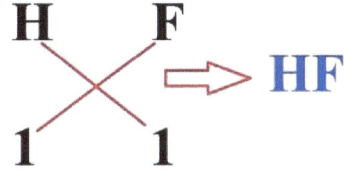

Formula   CS$_2$

Fig. 5 formation of CS2

Example 0.4

HF

Fig. 6 formation of HF

Example 0.5

Al_2O_3

Fig. 6 formation of Al2O3

SUPPLEMENTRY PROBLEME

MCQ

Q1. Nucleus consists of?

a) Electron and neutron

b) Electron and proton

c) Proton and neutron

d) None of these

Q2. Valency of oxygen is?

a) 2

b) -2

c) 1

d) None of these

Q3. $Na+Cl_2 \rightarrow ?$

What is the product of above reaction?

a) $NaCl_2$

b) $NaCl_3$

c) $NaCl$

d) None of these

Q4. How many elements has synthesized in lab till now?

a) 22

b) 26

c) 92

d) 118

Q5. Which is an inert gas?

a) Hydrogen

b) Nitrogen

c) Oxygen

d) Neon

Q6. Which one is the heaviest among these?

a) Electron

b) Proton

c) Neutron

d) All has same masses

Q7. Which one has independent existence in a chemical reaction?

a) Electron

b) Atom

c) Molecule

d) None of these

Q8. How many electrons are there in a Na+ ion?

a) 10

b) 11

c) 12

d) 9

Q9. Why does any atom react with other?

a) to give its electron

b) to gain electrons

c) to get a stable configuration.

d) none of these

Q10. Cl_2 is ?

a) Monatomic

b) diatomic

c) Inert gas

 d) none of these

Choose whether these statements are true or false

a) Electrons have negative charge.

b) In a simple hydrogen atom one electron and one proton is present.

c) Electrons revolve around the nucleus.

d) Electronic configuration is helpful in finding whether a given element is inert or not.

e) Chlorine has positive valency.

f) Hydrogen molecule is diatomic.

CHAPTER 1

STRUCTURE OF ATOM

What are insulators and conductors?

The materials which allow electricity to pass through them are known as conductor and those not known as insulator.

We can demonstrate it in a simple way. We break the wire of a simple closed circuit containing a d.c battery, bulb and connecting wire. Then join the material ends to the ends of wires. If bulb glows it's a conductor otherwise insulator.

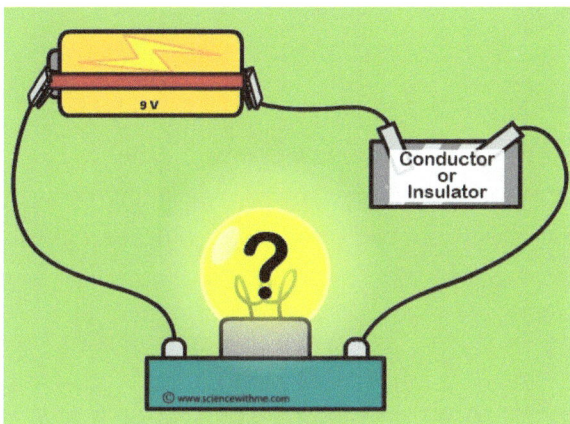

Fig. 7 activity to check insulator and conductor

How to demonstrate whether a liquid is insulator or not?

A liquid can be also demonstrated in the similar manner as stated above. Only slight change is that the liquid is kept in a beaker or insulated pot. The open ends of the wire are kept in the liquid and glowing of bulb is observed.

Is air insulator? How to demonstrate it?

If we take in the succession of above experiment, here we have to take a closed tube for gas or air taken under observation. In general it seems that air is insulator at normal temperature and pressure. It is right. Because if it was a conductor at room temp. and pressure our room had electrified by switch of electric board and we would not have survived as now.

But what at a very high voltage and low pressure supplied to the taken discharge tube. J.J Thomson raised the voltage more than 10000v in the similar experiment and reduces the pressure introducing a vacuum tube in the discharge tube. He observed the flickering yellow greenish light on the ZnS plate kept opposite of the cathode.

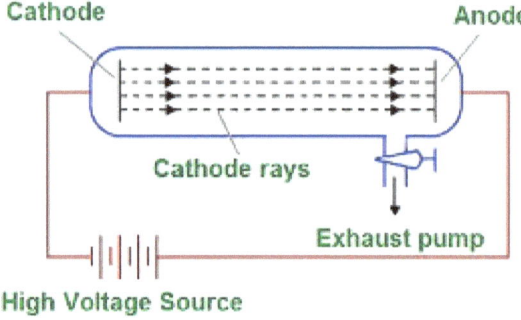

Fig. 8 vacuum tube in Thomson's experiment

It was concluded that the air is conductor at very high voltage and extremely low pressure. One important conclusion from this experiment leads to the discovery of electron. The ray which illuminated the ZnS screen was termed as the cathode ray as it was originated from cathode. It was composed of the negatively charged particle termed as **electron.**

Discovery of proton

At that time it was assumed that the atom as a whole is neutral. After the discovery of electron it was seen that on removal of electron the rest part must be positively charged. For this perforated cathode was taken and the screen was kept opposite the anode. The remaining +ve charged where moved towards the screen and seen as a fluorescent on the screen. They were passing through hole or canal of the cathode and termed as the **canal ray**. The constituent particle of the canal ray were examined and given the name proton.

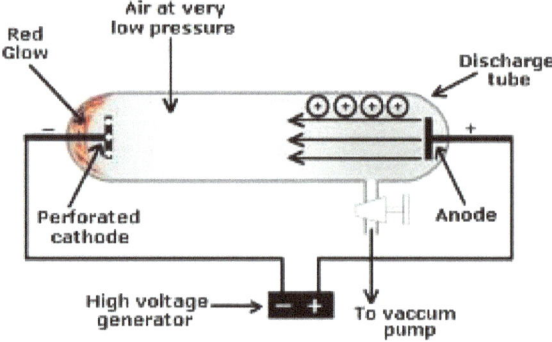

Fig. 9 observation of canal rays

Discovery of neutron

It was observed that the mass of atom other than electron was approximately double of the proton. So, there must be a particle having mass equal to the proton and neutral in charge inside the atom. The name of this particle was given by JAMES CHADWICK. It is now referred as neutron.

Image.1 James Chadwick

ATOMIC MODELS

After the discovery of electron, proton and neutron, it was a big challenge to represent the arrangement of these sub atomic particles inside the atom. First model was forwarded by J.J Thomson.

J.J Thomson model of atom

J.J Thomson gave the plum pudding modeling. According to him the atom was like a watermelon. The seeds were referred as the electrons embedded in the positively charged watermelon. The +ve and –ve charge were balanced to provide a neutral atom.

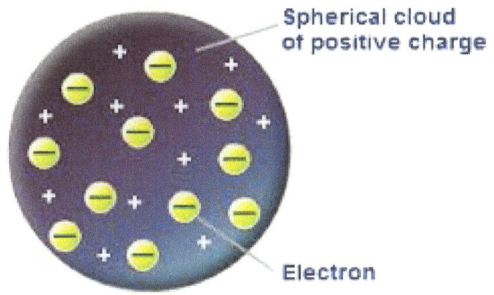

Thomson's Plum pudding model

Fig. 10 Thomson's model

Fig. 11 Thomson's theory

Rutherford's scattering experiment

Aim: earnest Rutherford was interested in finding the arrangement of the electrons inside the atom. For satisfying his curiosity he conducted an experiment.

Arrangement: He was interested in taking a very thin layer of atom. For this purpose he took a fine layer of gold. It was approximately 100 atoms thick. This thin layer was bombarded with alpha particle.

Observation:

a) Almost all alpha particles were going without any deviation.

b) Very few were deflected at small angle.

c) A single particle in 12000 approximately deflected at 180 degree.

Conclusion:

a) Most of the part inside the atom is empty.

b) As alpha particle is positively charged and approximately 8000 times heavier than electron but it is returning so almost all mass of atom and positively charges must reside at its centre.

c) Almost 99% alpha particle is going undeviated so size of nucleus is very small.

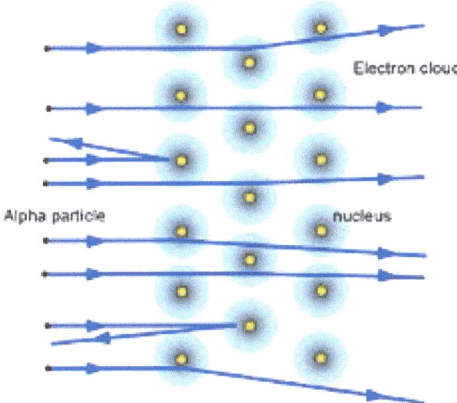

Fig. 12 scattering experiment

Rutherford's atomic model

As per the experimental result Rutherford gave the following atomic model

a) Atom is like a sphere and there is nucleus in its centre. Nucleus is positively charged and whole mass of atom resides inside nucleus.

b) Electrons revolve round the nucleus in circular orbit.

c) The size of nucleus is very small.

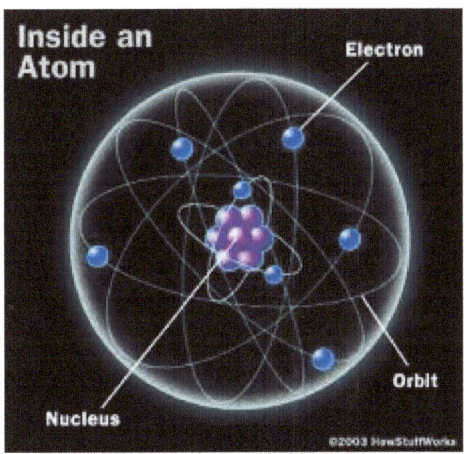

Fig. 13 Rutherford's model

Image.2 Ernest Rutherford

Drawback's of Rutherford's atomic model

According to the Rutherford's atomic model, electron revolves around the nucleus. As per electromagnetic theory, every moving object emits electromagnetic radiations and thus loses its energy. It is just like running in a circular park. If so happens electron must loses energy and spiral into the nucleus within a time span of 10^-6 second. In this case electron will not exist so atom and so this universe, but it is! Rutherford was unable to explain it. It was a great drawback of Rutherford's atomic model.

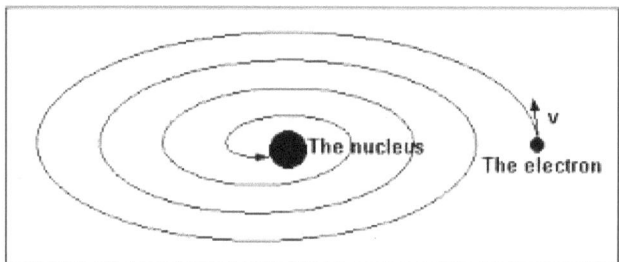

In the planetary model of atom, the electron should emit energy and spirally fall on the nucleus.

Fig.14 drawback of Rutherford's atomic model

Bohr's model of atom

Bohr's model resolves the drawback of Rutherford's atomic model. According to this model

a) Every electron moves in a fixed energy orbit.

b) They do not lose energy during their revolution.

Somerfield's atomic model

Somerfield explained that the electrons do not revolve in a circular orbit but elliptical.

SIZE OF ATOM AND NUCLEUS

In general form if we consider the nucleus as tip point of our pen's nib. The dimension of our room will be the atom. If we take a cricket ball as a nucleus then a circle of radius 5km will constitute the first orbit of the atom, taken ball as a centre. If we take nucleus as a centre of our solar system then in proportion our solar system will be smaller than the atom. That's why now a day it is an area of research and emergence of nanotechnology.

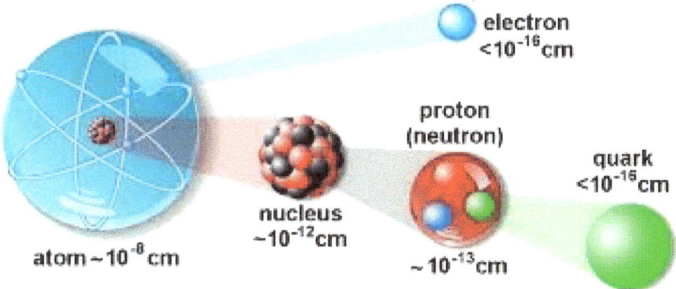

Fig. 15 size of atom and nucleus

ISOTOPE

There are many atoms found naturally which have same atomic no. but different atomic masses. These atoms are referred as the isotope of each other.

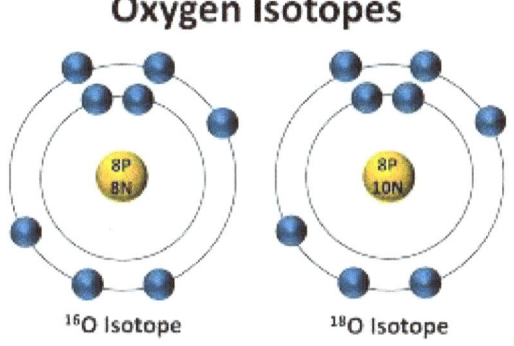

Fig. 16 oxygen isotopes

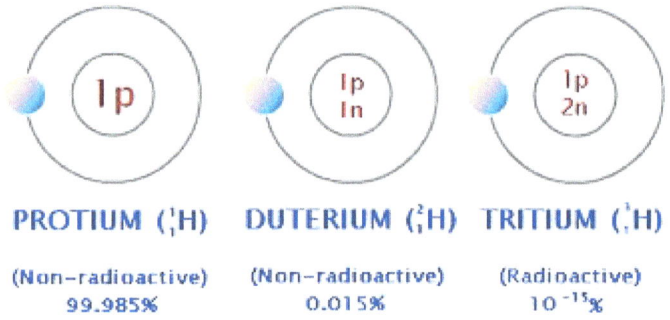

PROTIUM ($^1_1$H)   DUTERIUM ($^2_1$H)   TRITIUM ($^3_1$H)

(Non-radioactive)   (Non-radioactive)   (Radioactive)
99.985%   0.015% 10^{-15}%

Fig. 17 isotopes of hydrogen

ISOBAR

There are many atoms which have same no. of atomic masses but different atomic number. These atoms are referred as isobar of each other.

Example 1.1

(i) $_{32}Ce^{76}$, $_{34}Se^{76}$

(ii) $_{26}Fe^{58}$, $_{27}Ni^{58}$

ISOTONE

Atoms having same number of neutrons are known as isotones of each other

Example 1.2

Example:

Oxygen $^{16}_{8}O$ (p=8; n=8)

Nitrogen $^{15}_{7}N$ (p=7; n=8)

Carbon $^{14}_{6}C$ (p=6; n=8)

are isotones because of having same no. of neutron (8).

SUPPLEMENTRY PROBLEME

MCQ

Q1. Cathode ray consists of?

a) Electron

b) Atom

c) Molecule

d) None of these

Q2.canal ray consists of?

a) Electron

b) Atom

c) Molecule

d) None of these

Q3. Neutron has ....charges?

a) Negative

b) Positive

c) None of these

d) Some time negative some time positive

Q4. When cathode ray observed in J.J Thomson's experiment?

a) At high temperature and high pressure.

b) At very high voltage and very low pressure.

c) At high voltage and high pressure.

d) None of these

Q5. Who discovered electron?

a) J.J Thomson's

b) E. Goldstein

c) James Chadwick

d) Neils bohr

Q6. . Who discovered neutron?

a) Rutherford

b) E. Goldstein

c) James Chadwick

d) Neils bohr

Q7. Isotopes have same…………….?

a) Atomic masses

 b) no. of neutrons

c) Atomic number

d) none of these

Q8.Cl-35/17 and Cl-37/17 are?

a) Isotopes

b) Isobars

c) Isotones

d) None of these

Q9. N-14/7 and C-14/6 are?

a) Isotopes

b) Isobars

c) Isotones

d) None of these

Q10. In alpha particle scattering experiment Rutherford chosen?

a) Zn foils

b) Silver foil

c) Gold foil

 d) None of these

State whether these statements are true or false

a) Rutherford gave the plum pudding model of atom.

b) Somerfield explained that the electrons revolve in an elliptical orbit.

c) Isotones have same no. of protons.

d) The whole mass of atom is concentrated in the nucleus.

e) James Chadwick discovered the neutron.

f) Nucleus is neutral.

g) Atom is neutral but nucleus is negatively charged.

h) Isotope of iodine is used to cure goiter.

Short answer type questions:

Q1. Write two differences between canal ray and cathode ray.

Q2. What was the result of Rutherford's scattering experiment? Mention any two.

Q3.what are isotopes. Write three isotopes of hydrogen.

Long answer type questions:

Q1. Briefly explain the Thomson's model of atom.

Q2. Explain the alpha particle scattering experiment keeping in view following topics:

    a)   Aim   b)Observation   c) Conclusion

CHAPTER 2

ATOM AND MOLECULES

LAW OF CHEMICAL EQUATION

Molecules have independent existence. Only these particles actually take part in chemical equation. Atoms have no any independent existence. In every reaction the law of chemical equation is followed. Law of chemical equation consists of two basic laws which are discussed in the following sections:

LAW OF CONSTANT PROPORTION

In any chemical reaction the constituents' elements are in proportion of their masses. For example: if take water from different source like sea, tape water, formed in the lab, rain and stored in fields ,each will have the mass ration of hydrogen to oxygen as 1:8.

Oxygen	+ Hydrogen	→ Water	+ Oxygen	+ Hydrogen
8 g	1 g	9 g		
10 g	1 g	9 g	2 g	
8 g	2 g	9 g		1 g

leftover unreacted chemicals

Table 2

LAW OF CONSERVATION OF MASS

Law of conservation of mass states that in chemical reactions mass can neither be created nor be destroyed. In a given chemical reaction the mass of reactants as well as mass of product is conserved.

Example 2.1

$$e.g.\, HgO\,(s) \xrightarrow[close\ tube]{\quad\lrcorner\quad} Hg\,(\ell) + \frac{1}{2}\,O_2\,(g)$$

100g   92.6g   7.4g

Example 2.2

$$2H_2 + O_2 \rightarrow 2H_2O$$

4H,   2O =   4H, 2O

DALTONS ATOMIC THEORY

For the first time Dalton gave the term "atom". He suggested that the matter, whether it is element, compound or mixture consists of a small particle called atom. About atom he gave some postulates generally referred as Dalton's theory of atom. In a simple way we can understand them as follows:

a) Matter is made up of very small particle known as atom.

b) We can't create or destroy atom.

c) Atoms of a particular element are similar in chemical properties as well as atomic masses.

d) Atoms of different elements have different atomic masses as well as chemical properties.

e) In a compound the atoms combine in simple ratios of whole number.

f) In a compound the relative number and kinds of atoms is constant.

ATOMIC MASS

For a given atom the atomic no. refers to the sum of the no. of protons and the neutrons.

FORMULA MASS AND MOLECULAR MASS

Formula mass refers to the sum of the masses of the total atoms present in the given formula of a compound. Similarly the molecular mass is the sum of the masses of the atoms present in the given molecule.

CONCEPT OF MOLE

In the above sections we have studied that the size of atom and molecules. In real life if we take a half tea spoon of any chemical compound, we can't say how many atoms or molecule are present there? In general life we handle a large no. of atoms and molecule. In this situation the mole concept gives a handy rule to estimate the no. of atoms and molecule in a given quantity of chemical.

In a dozen of banana there are 12 bananas. In a pair of shoes there are 2 shoes. As dozen and pair refers to numbers 12 and 2 respectively. In a similar manner the mole refers to the number 6.022×10^{23}.

 But now the problem is finding how many moles are there in 4 gram of hydrogen atom. We can understand it in a easy way if we take the no. of gram equal to the atomic mass it will contain one mole. As hydrogen has atomic mass 1 amu so, 1g hydrogen constitutes 1mol of its atom. Similarly 4g of hydrogen atom will contain 4 mole of hydrogen.

Example 2.3 how many mole of H_2SO_4 molecule is present in 49 gram of Sulphuric acid.

Solution:

Step1:

Here, we have to find the mole of molecule. So, we will first find the molecular mass. The molecular mass of the H_2SO_4 is (2+32+64) 98u.

Step2:

Therefore 98 gram will constitute 1 mol of molecule.

Step3:

Therefore 49g will constitute (49/98) 0.5 mole.

Molarity

Molarity is used to express the concentration of any given solution.

Molarity = (no. of moles of solute present in solution)/(volume of solution)

SUPPLEMENTRY PROBLEME

Q1. The word atom was given by?

a) Bohr

b) Rutherford

c) Lavoisier

d) Dalton

Q2. Law of conservation of mass is given by?

a) Bohr

b) Rutherford

c) Lavoisier

d) Dalton

Q3. Molecular mass of CuSO4 is given what? Given (Cu=63u, S=32u, O=16u)

a) 157u

b) 159g

c) 159u

d) 1.59u

Q4. The number of entities in one mole is equal to?

a) 1.6×10^{-19}

b) 1.6×10^{-27}

c) 6.022×10^{23}

d) None of these

Q5. Total number of electrons present in 44g of CO_2 gas is?

a) 1 mol

b) 11 mol

c) 22mol

d) None of these

Q6. When 196g of HCl is dissolved in 5 litre water. What is the molarity of the solution?

a) 0.2 mol/litre

b) 0.4 mol/litre

c) 2 mol/litre

d) None of these

Q7. How many sodium ions (Na+) are there in a 58gram of NaCl? (given Na=23u, Cl=35u)

a) 2 mol/litre

 b) 1 mol/litre

c) 0.5 mol/litre

d) none of these

Short answer questions:

Q1. Briefly explain the following:

a) Law of conservation of mass

b) Law of constant proportion

c) Molar mass

Long answer questions:

Q1. What is Dalton's atomic theory? Explain its postulates.

CHAPTER 3

MATTER

Introduction

Anything which has mass and occupies space is known as matter. Almost everything surrounding us like air, water, books and food we eat are matter.

Is air matter?

Air is essential for our life. We can feel it. But a general quest is that is it matter? In a more formal way has it mass or volume? If we fill a balloon its volume increases so air has volume. A filled balloon has more mass then unfilled so it must have mass. As it has mass as well as volume air is a matter.

What constitute a matter?

As atom has no any free existence matter is considered as constitute of molecules.

Properties of matter constituents: the molecule.

The molecules of a matter have some special properties like:

a) They attract each other: In matter the molecules are attached to each other due to inter-molecular forces. In solid it is highest and lowest in gases.

b) There are spaces between them: In any matter there is space between the molecules. These spaces are termed as intermolecular spaces. In solids it is less and highest in gases.

It can be seen in an easy demonstration. When we make a solution of sugar and water, the sugar adjusts in the intermolecular spaces of the water and sugar is not seen in the solution.

c) They are continuously moving: whether it is solid, liquid or gases the molecules are in continuous motion.

For example, due to this property we are able to smell of a tasty food in other room.

States of matter

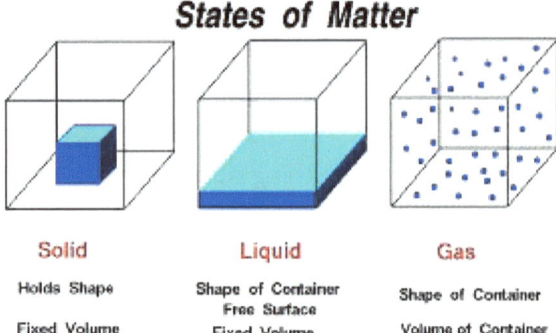

Fig. 18 states of matter

Solid

In solid the intermolecular force is maximum and the intermolecular spaces least. For this reason solids have fixed shape and volume.

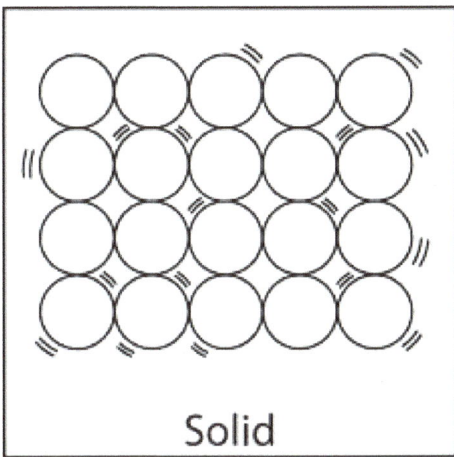

Fig. 19 molecules in solid

Liquid

In liquid the intermolecular force is less than solids but greater than liquids. The intermolecular spaces are more than solid but less than gases. For this reason liquid has fixed volume. This intermolecular force is not enough to fix the shape and size. Therefore the liquid takes the shape of the container in which it is kept.

Fig. 19 liquid has no shape

Gases

In the gases the intermolecular spaces are most and the intermolecular force is least. That's why molecules are free to move in entire space. Gases have no fixed volume nor shape and size.

Plasma

Plasma is the fourth state of matter in which the negative and positive ions are mixed and form the matter. These are loosely bonded in comparison to the gases. Approx charge on the plasma is zero.

Bose Einstein condensate (BEC)

The fifth state is Bose Einstein condensate. Its compactness is much more than a solid state. Generally it is the gases cooled at the 0 Kelvin.

Changing of states of matter

Solid, liquid and gases are transformed from one state to other due to change in temperature and pressure.

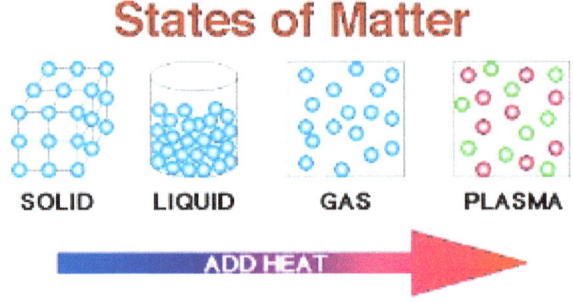

Fig. 20 states of matter

Melting: When heat is added to the solid it changes to the liquid. This process is known as melting.

Freezing: Freezing is the reverse process of the melting. In this process Liquid changes to the solid.

Condensing: In condensing process gases are changed to the solid on decrease of temperature.

Boiling: In boiling the liquid is converted into gaseous state on the account of heat gain.

Sublimation: Sublimation is the single process which is responsible for the inter conversion of the solid to gases and vice-versa.

Ionization: Gaseous state is converted to the plasma state when the gaseous atoms are ionized.

Deionization: Deionization is responsible for the conversion of plasma to the gaseous state.

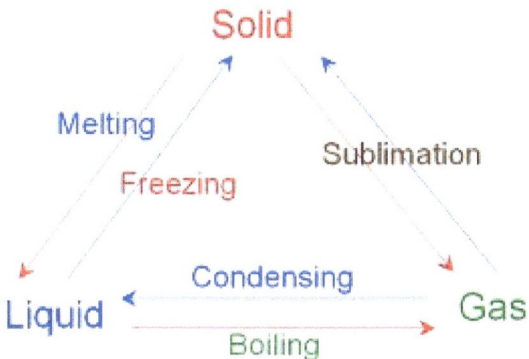

Fig. 21 changing the states of matter

Evaporation: a process of cooling

Definition: below the boiling point of any liquid, the process of converting liquid into vapor is known as evaporation.

In an open liquid surface, there are many particles or molecule which have enough energy to break the intermolecular forces and changes into vapor. This process is may be at any temperature and generally referred as the evaporation. It is a cooling process because in evaporation the lost energy of the liquid surface is gained from surrounding and the surrounding cools.

SUPPLEMENTRY PROBLEME

MCQ

Q1. Which has fixed shape and volume?

a) Solid

b) Liquid

c) Gas

d) Plasma

Q2. In which intermolecular space is Maximum?

a) Solid

b) Liquid

c) Gas

d) all has same

Q3.difussion is involved in?

a) Solid

b) Liquid

c) Gas

d) None of these

Q4. The process of changing solid to gas directly is known as?

a) Evaporation

b) Condensation

c) Sublimation

d) Melting

Q5. On adding heat the intermolecular forces between two molecules:

a) Increases

b) Decreases

c) First increases then decreases

d) None of these

Short answer questions:

Q1. Compare the properties of solid, liquid and gases.

Q2.What is the difference between melting and sublimation?

Q3. How can you demonstrate that "oxygen is matter"?

CHAPTER 4

PURE AND IMPURE MATTER

Introduction

Matters surrounding us can be separated in two groups i.e. pure and impure. Pure matters have fixed composition while impure have no fixed composition. Further pure substances can be divided into two group's elements and compounds. Elements cannot be further broken into simpler substances. Examples of elements are iron, copper, sodium, uranium etc. however compounds can be broken into simpler substances by chemical and electrochemical processes. Impure matter has no fixed composition however may have uniformity in distribution such mixture is termed as homogeneous mixture. Example of homogeneous is water in alcohol. In a general form we can say that the elements and compounds are pure matter and mixtures are impure.

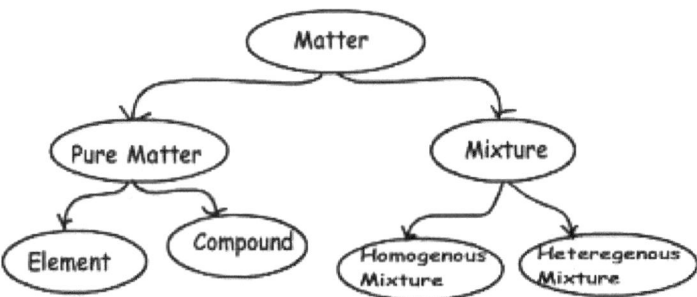

Fig. 22 classification of matter

Physical and chemical changes

Physical change only involves the change of states. In chemical change the property of the matter also changes and in addition heat and light evolves. Changing of ice into liquid is physical change while rusting of iron is chemical change.

Mixture

When two or more than two elements or compounds combine physically in any proportion mixture is formed.

Mixture can be homogeneous or heterogeneous. If the constituents of mixture are uniformly distributed over the entire range the mixture is called homogeneous mixture. Sugar solution is the example of homogeneous mixture. In heterogeneous mixture the distribution of constituent is non uniform. Mixture of sand and rice is the example of heterogeneous mixture.

Solution

Solution is a homogeneous mixture. It may be the mixture of two solids like alloy, two or more gases like air. It may also be the homogeneity of more than two states like sugar dissolved in water. The constituent which dissolves is known as solvent and which is dissolved called solute. Solutes can't be filtered out.

Properties of solution:

1. In a solution the solutes are of the order less than nano metre. So, we can't see them by our naked eye.

2. It can't scatter the light passing through them.

3. Solutes cannot settle down when kept the solution in undisturbed state.

Expressing concentration of solution:

(i) Mass percentage of a solution= (Mass of solute/Mass of solution)x100

(ii) Volume percentage of a solution= (Mass of solute/Volume of solution)x100

Suspension

A suspension is heterogeneous mixture. Its solute can be seen by naked eyes. It scatters the light and shows a visible path. it is not stable like solution because in undisturbed conditions its solutes settle down and the suspension behavior is suspended. Its solute can be separated by using filter.

Colloidal

In colloidal the size of solute is in between the solution and the suspension. A special name dispersed phase for solute and dispersing medium for solvent is used. The dispersed phases are such that they can't be filtered out. However dispersed phase can be separated using centrifugation technique. They also show the scattering of light. Colloids are named as per their phase and medium. Some examples are illustrated below.

Example of colloids:

Dispersed phase	Dispersing medium	type	example
Liquid	gas	Aerosol	Fog, clouds, mist
Solid	Gas	Aerosol	Smoke, automobile exhaust
Gas	Liquid	Foam	Shaving cream
Liquid	Liquid	Emulsion	Milk, face cream
Solid	Liquid	Sol	Milk of magnesia, mud
Gas	Solid	Foam	Foam, rubber, sponge, pumice
Liquid	Solid	Gel	Jelly, cheese, butter
Solid	Solid	Solid Sol	Coloured gemstone, milky glass

Compound

When two or more than two elements combine chemically in a definite proportion it results into a compound. For example $2H_2$ molecules combine with single oxygen molecule O_2 to form two molecule of H_2O.

Differences between mixture and compounds

Mixture

1. The resultant mixture is not a new compound.

2. It has none uniform composition.

3. The resultant mixture has the properties of the constituent element or compounds.

4. The components can be separated using simple physical methods.

Compound

1. The resultant is a new compound.

2. It has uniform and fixed composition.

3. The resultant compound has the different properties from its constituting elements.

4. Its components can be separated only using chemical or electrochemical processes.

Some methods to separate the components from mixture:

Evaporation

It is used to separate the mixture of volatile and non volatile substances.

Chromatography

It is used to separate the solutes which are dissolved in the same solvent.

Applications

To separate
a) Colours in a dy
b) Pigments from natural colours
c) Drugs from blood

Separating funnel

It is used to separate two immiscible liquids like water and oil.

Centrifugation

It is based on the principle that when a mixture of solid in liquid is spun rapidly the lighter one come on the top and denser to bottom.

Applications

a) In diagnostic laboratories for blood and urine tests.
b) In dairies and home to separate butter from cream.
c) In washing machines to squeeze out water from wet clothes.

Fractional distillation

It is used to separate the mixture of two miscible liquids. Petroleum products are separated from this technique. It is also useful in separating different gases from the air. For this purpose we decrease temperature and increase pressure to form a liquid air. Liquid air is then fractionally distilled to get the different gases.

Sublimation

It is used to separate solids from mixture which sublime i.e. ammonium chloride, camphor, naphthalene and anthracene.

Crystallization

It is the process of purifying solids.

Applications

a) Purification of sea salt.

b) Separation of crystals of alum from impure samples.

Q1. NaCl is a?

a) Pure substance

b) Impure substance

c) Mixture of pure and impure

d) None of these

Q2.solution of sugar in water is?

a) Element

b) Compound

c) Homogeneous mixture

d) Heterogeneous mixture

Q3. Milk is a?

a) Colloidal

b) Solution

c) Suspension

d) Compound

Q4. Cream is obtained from milk using which processes.

a) Filtration

b) Chromatography

c) Sublimation

d) Centrifugation

Q5. Which process is used to separate drugs from blood?

a) Filtration

b) Chromatography

c) Sublimation

d) Centrifugation

Q6. Which process is used to squeezes water from cloth in washing machine?

a) Filtration

b) Chromatography

c) Sublimation

d) Centrifugation

ANSWER TO THE SUPPLEMENTRY PROBLEMS

CHAPTER 0

1.C 2.B 3.C 4.C 5.D 6.C 7.C 8.A 9.C 10.B

TRUE   A, B, C, D, F,

FALSE   E

CHAPTER 1

1.A 2.D 3.C 4.B 5.A 6.C 7.C 8.A 9.B 10.C

TRUE   B, D, E, H

FALSE   A, C, F, G

CHAPTER 2

1.D 2.C 3.C 4.C 5.C 6.B 7.B

CHAPTER 3

1.A 2.C 3.C 4.C 5.B

CHAPTER 4

1.A 2.C 3.A 4.D 5.B 6.D